When Honey the Tiger Flew

Printed in China. Interior pages printed on FSC certified paper.

Written in Collaboration with the Somaly Mam Foundation Voices for Change Program Participants 2010

Editor: Nic Buckley and Anna Ludeke
Illustrations: Prak Ke
Photos: Somaly Mam Foundation, Wildlife Conservation Society, and Kevin McCaffrey
Resources: Wildlife Conservation Society
Map: Molly McCormick

Library of Congress Control Number: 2011930225
ISBN 13: 978-0-9842146-2-4
ISBN 10: 0-984214-62-3

From One of the Authors

Hello!

My name is Sina Vann and I am a survivor of modern-day slavery. At the age of 13, I was trafficked from my hometown in Vietnam to Cambodia, where I would suffer for many years before my hero, Somaly Mam, rescued me from my living nightmare. Over a decade later, I have become the Co-Director of a program called Voices for Change; a network of human trafficking survivors that supports and empowers young women and children who have been subjected to similar human rights violations. We are activists generating social change, while concurrently developing survivors into peer educators and rescuing victims from modern-day slavery.

Human trafficking is not only in Cambodia, but all over the world- including the United States. It will take all of us together, a united group of survivors and their peers, to stand up and stop the discrimination against the victims, to fight for human dignity, empowerment and opportunity for all young women and children around the world.

I hope you enjoy our story and will join us in our fight to create a better world for the next generation.

Love,

Sina and Voices for Change

Somaly Mam
Somaly Mam
Somaly Mam
Somaly Mam
Somaly.org
Somaly.org
Somaly.org
Somaly.org

About this Book

Hello Friends!

The young Cambodian women pictured to the left are the collaborating authors of this book. They are a part of a program called Voices for Change that provides survivors of modern-day slavery with a voice in their lives and a chance to support and empower others who have suffered the same terrible conditions. They are strong, resilient, courageous women who are building promising futures for themselves and countless other young women and children.

In preparing to work on this book, these young women spent a day learning about endangered Indochinese tigers. Then they used the lessons they learned to write stories about tigers individually and in groups. We combined our favorite parts of their stories into **When Honey the Tiger Flew**, adding some of our own ideas when needed to help the pieces fit together.

At the end of this book you can explore the world of tigers, discover the culture of Cambodia, and learn more about the authors. You can also read about two women who are working to make positive changes in the world. Somaly Mam, the founder of the Somaly Mam Foundation, has devoted her life to helping young women like the authors of this book overcome serious challenges and become leaders in Cambodia and the world. Hannah O'Kelly, a member of the Wildlife Conservation Society staff in Cambodia, is dedicated to preventing the extinction of tigers.

The team at Dot-to-Dot Children's Books hopes you enjoy **When Honey the Tiger Flew**. After reading this book, we invite you to write your own story about Indochinese tigers and to share it with us on our website at www.dot-to-dot-books.org.

In the mountains of Cambodia, deep in the green jungle, lived a young tiger named Honey.

He lived with his mother beside a river, just where it turned into a long, clear waterfall.

When Honey was a cub, his mom had taught him to hunt for sambar deer and wild pigs.

But each year, it was becoming harder and harder to find animals to hunt. Honey's mother grew more worried and weaker as she gave most of the food she caught to Honey. She had always been a strong and elegant hunter, but now was having such trouble finding food.

Honey wanted to help his beloved mother.
One morning, Honey was hungry and decided that he would do something himself to help. He set off through the forest hunting for a large animal to bring home to his mother.

But Honey could not find anything!
He searched and searched, and his stomach rumbled with hunger.
He even ventured close to a village, although he knew it was dangerous.

But all of the commotion of the village frightened Honey
and he retreated back into the deep jungle.

Frustrated and angry for not succeeding, Honey smashed his paw at the surface of a pond.

As the ripples settled, Honey saw that the pond was full of fish.
The fish looked fat and delicious. Very fat and delicious!

Honey had never been a good swimmer, even though his mother had tried to teach him. But now he thought excitedly, “Maybe I could catch a fish to eat!”

He was so hungry that he forgot his dislike of water
and plunged his paw into the lake and tried to grab one. SPLASH!

But by the time his paw reached the water,
the fish had darted away.

Honey waited until the bubbles cleared and the fish appeared once more. Again he raised his paw high above the water, and then swiped at an especially fat and delicious looking fish. SPLASH!

But the fish escaped his paw. He tried again and again until he was soaking wet—SPLASH! SPLASH! SPLASH!

Frustrated, hungry, and wet, Honey lay down and stared out over the water.

Just as he did, a great bird dove straight down from the sky and into the pond.

With a quiet graceful dive, the bird swooped underwater and flew away carrying a fat fish in its beak.

The fish flopped around wildly trying to escape, but the bird held him tight before gobbling him up.

Honey watched the bird jealously. That was his fish!
Why was the bird able to catch a fish and he wasn't?

Honey looked at his wet paws and then looked at the tree.
Honey had an idea—he could do just as the bird did! He would fly!

Finding a tree with low hanging branches,
Honey crept carefully out onto a branch over the water.
He crouched quietly and waited for the biggest fattest fish to swim by.

When it did, he closed his eyes and jumped
from the branch with all his might.

SPLASH! Honey plunged underwater, but there was no fish in sight. He had missed it!

Honey splashed about but was finally able to swim to the shore. Exhausted and dripping wet, Honey sat and thought once more of the fat delicious fish and his hungry mother.

Honey was tired, but looked back at the tree. No, he couldn't give up.

Taking a deep breath, he tiptoed onto the branch, determined to succeed.
He watched for the fish and closed his eyes and jumped in again.

SPLASH! He felt a fish between his jaws as he plunged into the water.

Again it slipped away! Discouraged and upset, Honey paddled back to shore.

"I must try one last time," he thought.

Crouching on the branch,
Honey focused carefully
on the fish below.

“Wait for the right moment,”
he thought quietly.

Wait . . . Wait . . . and . . .

JUMP!!! Honey leapt in with a great splash,
this time with his eyes wide open!

As he plunged under the water, he grabbed
the fish between his teeth and bit down sharply—and he caught it!

With the fish in his mouth he swam quickly to the shore, and shook the water from his fur. He felt so proud!

Honey was so happy, he wanted to show his mother. He raced home and laid the fish at his mother's feet.

"Honey!" she growled happily. "How did you catch this BIG fish? You hate swimming!"
"I flew!" he said. "But, Honey, tigers don't fly," she said.
"I'll show you tomorrow!" he replied.

That night in the forest, Honey and his mother ate the fat, delicious fish.
Honey felt proud of what he had accomplished.
He was turning into a good hunter, just like his mom.

Hi Friends!

We hope you enjoyed this story. People from all over the world helped to bring it to life. But the story isn't over yet. You are now a part of it and we need your help.

Although I spend most of my time by myself, when I am with my friends, we form a group called a streak. My friends at Dot-to-Dot Children's Books, Somaly Mam Foundation, Facing the Future, and the Wildlife Conservation Society have joined my streak and I hope that you will do the same.

You can join my streak by sharing the message of the Endangered Species, Empowered Communities (ESEC) stories with your friends and teachers. Whatever you do, taking action based on what you have learned about endangered species like me makes you a part of my team.

In the pages that follow this invitation, you can learn more about me and my friends. Just like the young women who helped to write this book, you can use what you learn to write your own story about tigers and then share it with us at www.dot-to-dot-books.org.

I hope to hear from you soon!
Honey the Tiger

Thank You!

My friends at Dot-to-Dot Children's Books and I want to express our sincere gratitude to the many people who helped us produce this book. First and foremost, we want to recognize the creative contributions of the young women who inspired and co-authored it.

I also want to thank my other friends from around the world who helped make the story come to life: Lisa Biasillo, Vanessa Birch, Mark Gately, Rob Gerstein, John Hendrix, Bill Livermore, Ed Pereira, Matthew Shea, Charlotte Salasky, Karen Smith, David Soman, Lisa Yook and Jan Bachman of Devenish Press.

Special thanks to Prak Ke and the team at Sonleuk Thmey – Marie, Anna, Borin, and Sokhom – for bringing **When Honey the Tiger Flew** to life with such beautiful illustrations. Prak Ke is a very talented Cambodian illustrator and Sonleuk Thmey is a Cambodian nonprofit organization that promotes literacy. To learn more, please visit our website.

Many thanks to everyone! I appreciate everything that everyone has done for me!

Your Friend,
Honey the Tiger

Indochinese Tigers

Indochinese Tiger Facts
Life Span: 15 - 20 years
Weight & Size: males are larger than females weighing up to 400 lbs, and measuring around 9 feet long
Diet: carnivore - wild pigs, deer, cattle
Habitat: forests and hilly, mountaineous terrain

Tigers are shy and nocturnal animals. When a tiger is relaxed, its tail hangs loosely but it twitches when the tiger is nervous or agitated.

Tigers also have glands in their paws that contain oils that other tigers can smell. Tigers mark their territory by scratching the bark of trees and leaving their scent. Marking territory lets other tigers know to keep out.

Tigers have adapted to their environment to help them hide from predators and prey. Their orange and white coats with black stripes blend into the forest around them. Their agility, speed, big claws, and teeth help make them great hunters.

The Indochinese tiger is one of only five tiger species left in the world today. Sadly, all tigers are endangered.

Endangered: Indochinese Tigers

Humans are the greatest threat to the Indochinese tiger. Cutting down trees for wood and to make room for human development has become a big problem for tigers because it has made their territories smaller and smaller. As forests are cleared for development, tigers lose their habitat along with the food sources that they usually hunt. Because of this, tigers are forced to hunt in different ways to find food, even fish, just like Honey in this story.

Tigers are also hunted by humans for their paws, fur, and bones. Some people believe that parts of the tiger can be used as medicine. They are illegally hunted and sold all over the world. This type of illegal hunting is called poaching and, unfortunately, it is a major reason for the decline of the tiger population. There may only be a few hundred Indochinese tigers left in the wild today. Protecting tigers through education and conservation is very important to avoid the extinction of this majestic animal.

The Wildlife Conservation Society (WCS) works with the government of Cambodia to ensure the protection of tiger habitat and prey species in its eastern regions. This area has been highlighted as the key habitat for tigers in Cambodia. WCS efforts specifically aim to reduce threats to tigers by preventing poaching and the illegal clearing of forest.

To learn more about the Wildlife Conservation Society and to support its efforts to protect wildlife and wild places worldwide, please visit www.wcs.org.

Meet One of My Friends

Human economic development across Asia is bringing serious challenges to the conservation of lands that are ideal habitat for tigers. Despite the hard work of conservationists in places like Cambodia, tigers are in danger of becoming extinct today. Hannah O'Kelly is one of these devoted conservationists.

Hannah grew up in Ireland but for the last five years she has been working on the other side of the world in Eastern Cambodia. She spends most of her time deep in the forest searching for tigers and other wildlife. Hannah came to Cambodia because it is one of the best places on Earth to study large mammals like tigers, leopards, Asian elephants, wild oxen, and many species of primates.

Hannah's job at the Wildlife Conservation Society (WCS) involves assisting her Cambodian colleagues in the research of various wildlife species, including tigers. Because tigers are extremely secretive and shy, it is difficult to know exactly how many tigers remain in Cambodia today and hunting is driving the tiger population toward extinction. To find tigers, Hannah uses a specially trained dog, named Maggie, to help her. Dogs have a better sense of smell than people and Maggie is able to sniff out tigers in the forest as Hannah and her team track tigers.

Hannah believes that all people should help protect and conserve endangered species. She also believes that the best way to do this is to learn all you can about tigers and to teach others about them.

Where is Cambodia?

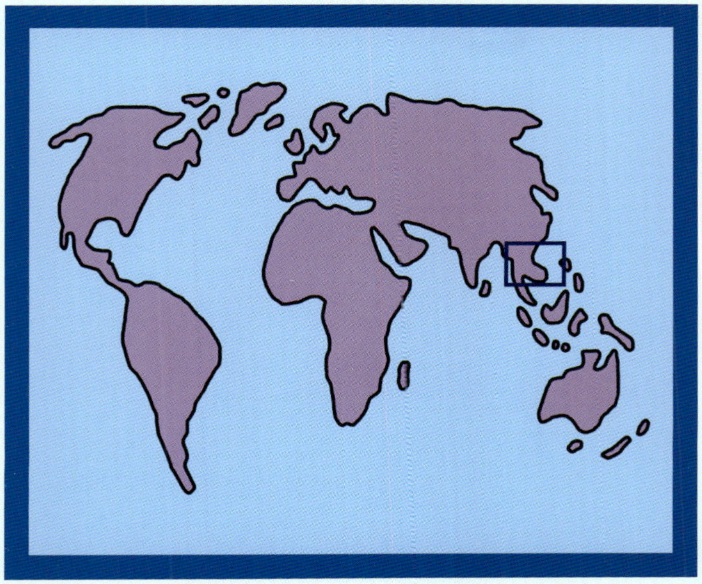

Cambodia Facts
Continent: Asia
Capital City: Phnom Penh
Official Language: Khmer
Neighboring Countries: Laos, Thailand, Vietnam

• Cambodia is a tropical country and this helps the nation grow its most important crop, rice.

• Cambodia's official religion is Buddhism. Buddhism is also the religion of about 6% of the world's population.

• Despite having once been a part of the colony French Indochina, Cambodia has been able to maintain its unique identity while adapting aspects of French culture.

Life in Cambodia

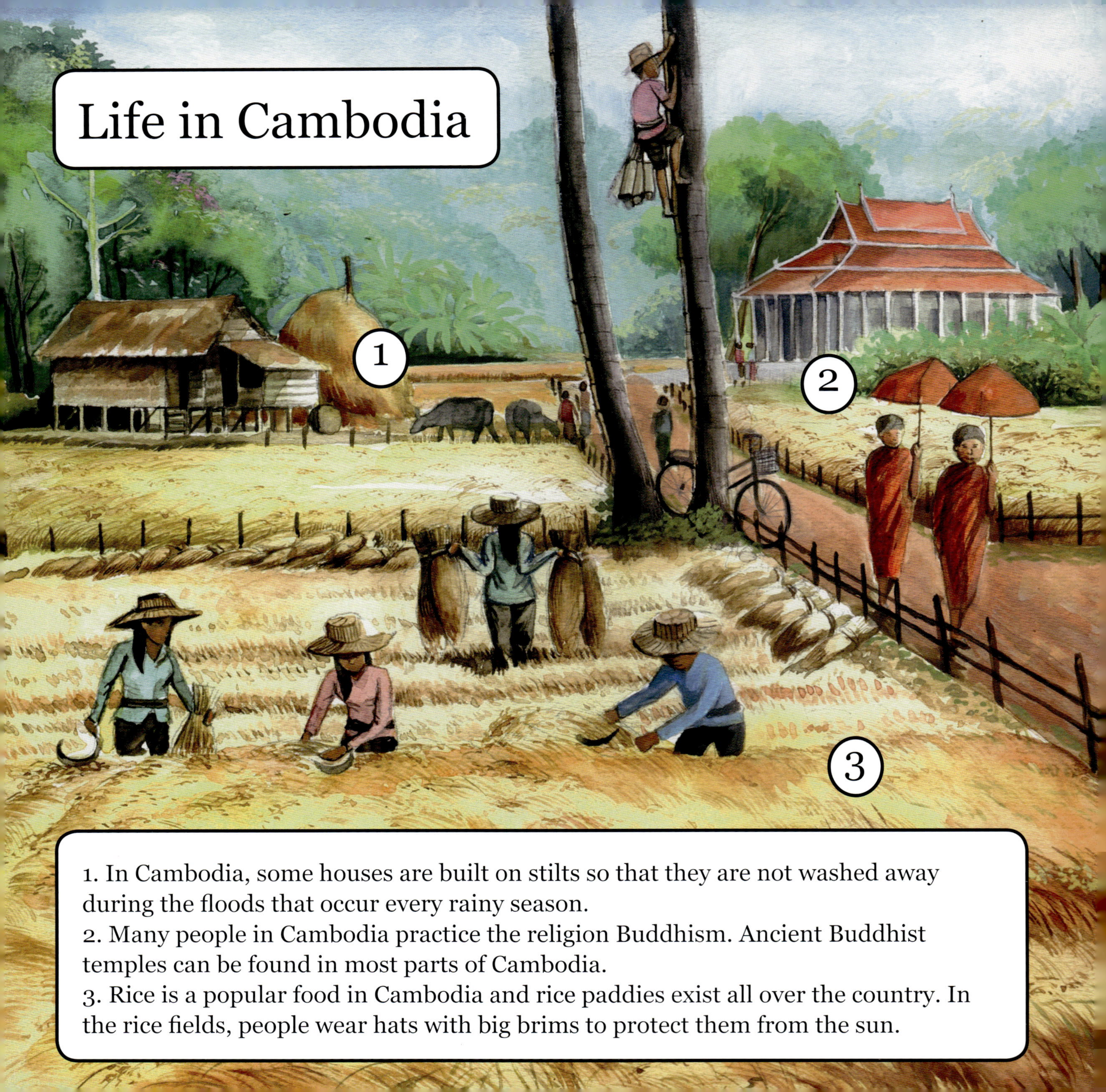

1. In Cambodia, some houses are built on stilts so that they are not washed away during the floods that occur every rainy season.
2. Many people in Cambodia practice the religion Buddhism. Ancient Buddhist temples can be found in most parts of Cambodia.
3. Rice is a popular food in Cambodia and rice paddies exist all over the country. In the rice fields, people wear hats with big brims to protect them from the sun.

Meet the Authors

The eleven young women who helped author **When Honey the Tiger Flew** are a part of Somaly Mam's Voices for Change Program. This program is a platform for survivors of slavery to become leaders and advocates for the anti-trafficking movement. They help rescue women and children from slavery, teach people about human trafficking, and serve as positive examples for thousands of other young people in Cambodia and around the world. These women fulfill the vision that from those who have struggled through the pain of slavery will arise a new generation of leaders who stand for justice and free will.

Y Li Mey would like to visit the ancient temples made by her ancestors at Angkor Wat. She is learning administrative skills to help her find a good job.

Srey Sokkhy recently finished an accounting internship with the Somaly Mam Foundation. She hopes to own her own restaurant in the future.

Thy Sopheap wants to start her own tailoring shop and run it with her daughters. She also wants to study psychology and help educate people in her community about slavery.

Chan Srey Pov is a proud Cambodian who enjoys studying English. In the future, she wants to be a lawyer or public speaker so that she can help the victims of slavery.

Srey Pich loves studying topics related to medicine and hopes to become a doctor one day. More than anything else, she wants to help people using her interest in medicine.

Him Chay Voeurn is working to become a tailor. Eventually, she hopes to own a shop and hire girls who have survived slavery.

Long Somana dreams about helping people who are at risk of becoming victims of slavery through education. She enjoys studying Khmer and English literature.

Meas Srey An is an avid reader and loves books. Her vision is to start a weaving business. She wants to train poor people in her business to help them improve their lives.

Man Sina wants to live in the countryside and have her own resort and a farm where survivors of slavery can work. She wants to reduce poverty by training people to farm more productively.

Yok Sotheara loves to draw as often as she can because it helps her release her stress. She is working toward the goal of having her own tailoring shop in the future.

Sina Vann, who wrote the letter at the beginning of this book, is the leader of the Voices for Change Program and a member of the Somaly Mam Foundation Board of Directors.

About the Somaly Mam Foundation

The Somaly Mam Foundation is a 501(c)(3) non-profit organization committed to empowering victims and survivors of human trafficking in Southeast Asia and around the world, and Dot-to-Dot Children's Books is very proud to highlight their work. Their vision is simple but essential: a world where women and children are safe from slavery.

You may be surprised to hear this: slavery and human trafficking are global problems today. Over two million women and children are sold into slavery each year. This is a crisis that must be stopped, but more resources are needed to prevent this atrocity and to support the rescue and recovery of victims.

The Somaly Mam Foundation is working to end human trafficking, liberate victims and empower survivors so they can live lives of dignity. Each day, they are helping to lead the fight against slavery around the world and each day they are making progress. The Foundation's work includes awareness and advocacy campaigns against human trafficking as well as creating educational opportunities for young women and girls.

Through its programs, the Somaly Mam Foundation helps survivors of slavery develop the identity and life skills to rebuild their lives. By learning how to speak English and how to use computers, the young women in the foundation's program gain valuable skills that will help them support themselves. Because of the Somaly Mam Foundation, former slaves have their freedom, a voice, and the chance to begin new lives.

To learn more about the Somaly Mam Foundation's work and impact, please visit www.somaly.org. Continue reading to meet the founder, Somaly Mam.

A Heroine and Leader

Meet Somaly Mam, Founder of the Somaly Mam Foundation

When one person decides to make a difference, the impact can be amazing. Somaly Mam was born into extreme poverty in Cambodia and kidnapped into slavery as a child. When she finally escaped years later, she did much more than take back her own life, she gave life back to thousands of other women who had endured the same horrific experiences.

Since gaining her freedom, Somaly has dedicated her life to fighting slavery every day. How? She started a nonprofit organization to help victims develop the physical, emotional, and mental strength to face the future with hope. She instills young women and children with the necessary knowledge and resources to speak out - to have a voice in their lives and control over their future. And because the power of many people working together is greater than that of a few, Somaly travels the world inspiring both advocates and survivors to speak out against human trafficking. She wrote a book, **The Road of Lost Innocence**, to tell her story and those of other survivors, and continues to put herself in danger to rescue others.

For her tireless efforts, Somaly has won worldwide respect and is known as a leader of the anti-trafficking struggle. In 2009, she was honored as one of Time Magazine's 100 Most Influential People and was featured as a CNN Hero. Somaly is a visionary force for positive change – dignified and strong.

Today, Somaly continues her courageous work in Cambodia. She lives among the women and children whom she rescues and stays by their side as they walk the difficult path to recovery and freedom.

We Need Your Help

In 2007, Kevin McCaffrey read an article about violence against women in Eastern DR Congo. Outraged and tired of inaction, Kevin decided to travel to Eastern Congo to meet survivors of the region's violence. During a performance of skits by some women survivors, Kevin was pleasantly surprised to find that they were gifted storytellers. Their creativity sparked an idea that would become Dot-to-Dot Children's Books.

Transformed by his experience, Kevin asked a few friends and one of his former teachers to help him found Dot-to-Dot. Although it was initially just a small group of friends, Dot-to-Dot did not remain small for long. Nearly 225 young women and children from eight countries joined the Endangered Species, Empowered Communities (ESEC) project by writing stories about endangered animals in their countries. Their writing was the inspiration for the dozens of volunteers who helped transform their stories into the children's books that are at the core of the ESEC project.

Dot-to-Dot Children's Books is a nonprofit organization that helps people around the world use their creativity to strengthen their communities while raising awareness about important global issues. Dot-to-Dot is also an example of the power of people joining together to support a common cause. "You want students to become citizens who work to solve real-world problems." says co-founder Paul Kim. "To work with them to publish books that make a positive impact is an amazing experience for me as a teacher."

Yet the ESEC project does not end with the publication of these books. The books themselves are simply a place to start. The true value of the ESEC project is only unlocked when readers are inspired to action by what they have read and learned. This is where we need your help. We hope you have been inspired to become a part of the ESEC story by taking action. Only through your actions can the ESEC project fulfill its potential.

What You Can Do

Spread the Word

Recommend the ESEC books to your friends and family, to local libraries, and to elementary school teachers. Each book sale will generate funding to help empower girls and protect endangered species while also raising awareness about the important issues featured in each book. Visit www.dot-to-dot-books.org to see details about how the funds will be used.

Share Your Creativity

Use your creativity to write stories and develop ideas that further the message of the ESEC project and share them with others on our website. Teachers can submit classroom materials, parents can recommend activities they have done at home, and kids can upload stories and illustrations they have created to share with others. All of these creative contributions will be available to you for free on our website along with some of our own suggestions about what you can do to help change the world.

Take Action

We hope that everyone who reads this book will commit to doing something - big or small - to help protect endangered species and empower communities in the world. Go to our website to find ideas about what you can do. Many of these ideas are simple and free.

For Educators

Connecting Students to the World: Engaging K-3 Lessons for Literacy, Science, and Social Studies from Facing the Future is a teacher lesson plan guide for grades K-3 designed to complement the ESEC books. Engage students with interdisciplinary lessons aligned with language arts, science, and social studies standards. As they explore real-world environmental and social issues, students discover how they can be involved in solutions. Download Connecting Students to the World at ww.facingthefuture.org.

Featured Organizations

Dot-to-Dot Children's Books is a not-for-profit organization that develops and publishes stories to further global education and to support community development around the world. As part of our mission, we want to raise awareness about the meaningful work of the organizations that are featured in this book.

Please join us in our proud support of their work.

Facing THE Future™

Facing the Future (FTF) believes in the transformative power of education to improve lives and communities, both locally and globally. FTF engages students by making academics relevant to their lives and by empowering them to think critically, develop a global perspective, and participate in positive solutions for a sustainable future. FTF's standards-based hands-on curriculum resources are in use in all 50 U.S. states and 120 countries. Learn more at www.facingthefuture.org.

Somaly Mam Foundation

The Somaly Mam Foundation (SMF) is dedicated to ending the trafficking and enslavement of women and children around the world. Each year, two million women and children become victims of this unspeakable crime. SMF fights against this epidemic with action, advocacy, and awareness programs that are inspired by the collective voice of survivors who have been able to reclaim their lives. Learn more at www.somaly.org.

WILDLIFE CONSERVATION SOCIETY

The Wildlife Conservation Society (WCS) saves wildlife and wild places worldwide. WCS does this through science, global conservation, education and the management of the world's largest system of urban wildlife parks, led by the flagship Bronx Zoo. In addition, WCS runs the New York Aquarium, Central Park Zoo, Prospect Park Zoo, and Queens Zoo. Together these activities change attitudes toward nature and help people imagine wildlife and humans living in harmony. WCS is committed to this mission because it is essential to the integrity of life on Earth. Teachers can find resources at www.bronxzoo.com/educators.aspx. To learn more about the Wildlife Conservation Society, visit www.wcs.org.